Guide
for the
Total Solar Eclipse
of
2026

Color Edition

Fred Espenak

Edition 1.0
May 2024

Guide for the Total Solar Eclipse of 2026 – Color Edition

Copyright © 2024 by Fred Espenak

All rights reserved. No part of this publication may be reproduced, distributed, or transmitted in any form or by any means including photocopying, recording, or other electronic or mechanical methods, without the prior written permission of the publisher, except in the case of brief quotations embodied in critical reviews and certain other noncommercial uses permitted by copyright law. For permission requests, write to the publisher at the address below.

Astropixels Publishing
P.O. Box 16197
Portal, AZ 85632

Astropixels Publishing Website: *www.astropixels.com/pubs*

This book may be ordered at: *www.astropixels.com/pubs/Guide2026.html*

Printed in the United States of America

ISBN 978-1-941983-46-1

Astropixels Publication: AP042 (Version 1.0a)

First Edition

Front Cover: A map of the path of the 2026 total solar eclipse through Greenland, Iceland and Spain. Map copyright © 2024 by Fred Espenak. More about the eclipse can be found at:

www.eclipsewise.com/solar/SEnews/TSE2026.html

Back Cover Photo of Fred Espenak: Copyright © 2024 by Patricia T. Espenak

Table of Contents

1 – INTRODUCTION 5
2 – ORTHOGRAPHIC PROJECTION MAP OF THE 2026 TOTAL SOLAR ECLIPSE PATH 7
3 – OVERVIEW MAP OF THE PATH OF TOTALITY 7
    OVERVIEW MAP: PATH OF TOTALITY – RUSSIA, GREENLAND, ICELAND, AND SPAIN 8
4 – PATH OF TOTALITY 9
    TABLE 1 – PATH OF TOTALITY 11
5 – LOCAL CIRCUMSTANCES 11
    TABLE 2 - LOCAL CIRCUMSTANCES FOR THE TOTAL ECLIPSE IN ICELAND 12
    TABLE 3 - LOCAL CIRCUMSTANCES FOR THE TOTAL ECLIPSE IN SPAIN 12
    TABLE 4 - LOCAL CIRCUMSTANCES FOR THE PARTIAL ECLIPSE IN EUROPE 13
    TABLE 5 - LOCAL CIRCUMSTANCES FOR THE PARTIAL ECLIPSE IN AFRICA 14
    TABLE 6 - LOCAL CIRCUMSTANCES FOR THE PARTIAL ECLIPSE IN THE USA 14
    TABLE 7 - LOCAL CIRCUMSTANCES FOR THE PARTIAL ECLIPSE IN CANADA 15
6 – WEATHER PROSPECTS FOR THE 2026 TOTAL SOLAR ECLIPSE 16
    RUSSIA/SIBERIA 16
    GREENLAND 16
    ICELAND 17
    SPAIN AND THE BALEARIC ISLANDS 18
    CLOUDS AND WEATHER IN SPAIN 19
7 – DETAILED MAPS OF THE PATH OF TOTALITY 21
8 – ECLIPSEWISE.COM WEB SITE 21
    ECLIPSEWISE.COM AND THE 2026 ECLIPSE 22
9 – ECLIPSE PREDICTIONS 22
10 – TOTAL SOLAR ECLIPSES: 2026 – 2040 23
    TABLE 8 – DATA FOR TOTAL SOLAR ECLIPSES 23

DETAILED ECLIPSE MAPS 25
    MAP 01: RUSSIA 25
    MAP 02: GREENLAND (NORTHERN) 26
    MAP 03: GREENLAND (CENTRAL) 27
    MAP 04: GREENLAND (SOUTHERN) 28
    MAP 05: ICELAND 29
    MAP 06: SPAIN – 1 30
    MAP 07: SPAIN – 2 31
    MAP 08: SPAIN – 3 32
    MAP 09: SPAIN – 4 33
    MAP 10: SPAIN – 5 34
    MAP 11: SPAIN – 6 35
    MAP 12: SPAIN – 7 36
    MAP 13: SPAIN – IBIZA AND MALLORCA 37
    MAP 14: SPAIN – MALLORCA AND MENORCA 38

Photo 1 – The Sun's corona is only visible to the naked eye for a few minutes during a total eclipse of the Sun. This image was captured during the total solar eclipse of April 20, 2023 from western Australia.
© 2023 F. Espenak, www.MrEclipse.com

Photo 2 – The The Diamond Ring effect is seen just as totality ends. Captured during the total solar eclipse of April 8, 2024 from Mazatlan, Mexico.
© 2024 F. Espenak, www.MrEclipse.com

Figure 1 – The path of totality of the Total Solar Eclipse of 2026 begins in Russia and passes close to the North Pole. It then swings south though eastern Greenland, western Iceland, and northern Spain.

1 – Introduction

On Wednesday, August 12, 2026, a total eclipse of the Sun will be visible from the Northern Hemisphere. Although a partial eclipse will be seen from much of North America and Western Europe, the total phase in which the Moon completely covers the Sun (known as *totality*) will only be seen from within the narrow path of the Moon's umbral shadow as it sweeps cross the Arctic, Greenland, Iceland, and Spain.

It is only during *totality* that the Sun's faint and exquisitely beautiful outer atmosphere – the solar corona – is revealed to the naked eye, the landscape is plunged into an eerie twilight, and brighter stars and planets appear. The spectacle of a total solar eclipse is one of the most remarkable in nature. But to witness it you *must* be in the *path of totality*.

The course of the Moon's shadow begins in the eastern Russia and travels northward across the Arctic Ocean. The path of totality passes within 64 miles (103 km) of the North Pole before curving to the southwest. By the time the path reaches Greenland, it is traveling due south and crosses along the fjords of the island's eastern coastline. Slowly curving to the southeast, the path cuts across western Iceland and proceeds through northern Spain (barely clipping northeastern Portugal). The path finally ends at sunset in the western Mediterranean Sea after encountering the Balearic Islands (Ibiza, Mallorca, and Menorca).

The duration of totality also varies along the path being longest near Iceland (2 minutes 18 seconds) and shortest in the Mediterranean Sea (1 minute 33 seconds). These durations are for the middle of the path along the *central line*. As you move away from the *central line* (either north or south), the duration of *totality* decreases. It happens very slowly at first but then drops rapidly to *zero* as you reach the northern or southern edges or limits of the *path of totality*.

A partial eclipse will be visible from a much larger region including western Europe, north Africa, Canada, and northeastern USA. Eclipse glasses or some type of eye protection must be used for viewing the partial eclipse. Of course, those in the path of the total eclipse must remove protective glasses to witness totality itself.

This guide is designed to assist observers in locating and reaching the central eclipse path to enjoy the ineffable experience of totality.

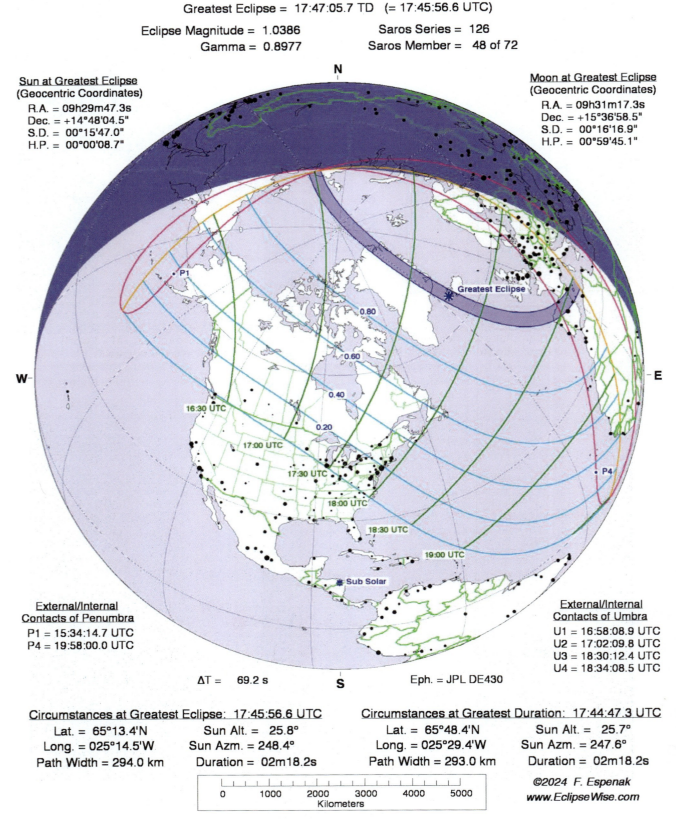

Figure 2 – Orthographic projection map of the Total Solar Eclipse of 2026.
Based on map from **21st Century Canon of Solar Eclipses** by Fred Espenak (2017)

2 – Orthographic Projection Map of the 2026 Total Solar Eclipse Path

The orthographic projection map of Earth in Figure 2 shows the path of penumbral (partial) and umbral (total) shadows of the Moon during the eclipse.

The limits of the Moon's penumbral shadow define the region of visibility of the partial eclipse. It covers about half of Earth's daylight hemisphere and consists of several distinct zones or limits. At the northern and southern boundaries lie the limits of the penumbra's path. Great loops at the western and eastern extremes of the penumbra's path identify the areas where the eclipse begins and ends at sunrise and sunset, respectively.

Bisecting the *eclipse begins and ends at sunrise and sunset* loops is the curve of maximum eclipse at sunrise (western loop) and sunset (eastern loop). Points *P1* and *P4* mark the coordinates where the penumbral shadow first contacts (partial eclipse begins) and last contacts (partial eclipse ends) Earth's surface. The path of the umbral shadow (*path of totality*) appears within the penumbral path. Curves of constant eclipse magnitude define points where the maximum magnitude has values of 0.2, 0.4, 0.6, and 0.8.

Above the map is the instant of greatest eclipse, expressed in Terrestrial Time[1] (TT) and Coordinated Universal Time[2] (UTC). The *Eclipse Magnitude* is the geocentric ratio of diameters of the Moon and the Sun. *Gamma* is the minimum distance of the Moon's shadow axis from Earth's center in Earth radii at greatest eclipse. Finally, the *Saros Series* of the eclipse is listed, followed by the *Saros Member* position. The first number identifies the sequence position of the eclipse in the Saros, while the second is the total number of eclipses in the series.

In the upper left and right corners are the geocentric coordinates of the Sun and the Moon, respectively, at the instant of greatest eclipse. They are:

                        R.A. – Right Ascension
                        Dec. – Declination
                        S.D. – Apparent Semi-Diameter
                        H.P. – Horizontal Parallax

To the lower left are the contact times of the Moon's penumbral shadow with Earth. They mark the times when the partial eclipse begins and ends, globally. To the lower left are the contact times of the Moon's penumbral shadow with Earth. They mark the times when the total eclipse path begins and ends, globally.

At bottom left are the geographic coordinates of the position of greatest eclipse along with the local circumstances at that location (i.e., Sun altitude, Sun azimuth, path width and duration of totality).

At bottom right are the corresponding values for the position of greatest duration. Although the geographic locations of greatest duration and greatest eclipse differ by about 70 km (43 miles) along the eclipse track, the difference in the duration of totality is less than 0.05 second.

3 – Overview Map of the Path of Totality

The overview map in Figure 3 focuses in on the path the total eclipse through Russia (shown in insert), Greenland, Iceland, and Spain. The white lines running across eclipse path mark the position of mid-eclipse at 10-minute intervals. The times are given in local time at each position along with the duration of totality on the central line, and the altitude of the Sun above the horizon at that instant.

The Mercator projection used in this map preserves local directions and shapes at the expense of inflating the size of objects the further they are from the equator. This is why Greenland looks so large when it is actually smaller than Europe.

---

[1] Terrestrial Time or TT is a time scale established by adopted by astronomers to calculate the location or planetary position of other objects as they relate to the center of the Earth and its rotation.

[2] Coordinated Universal Time or UTC is the primary time standard that regulates world clocks and time. It is based on International Atomic Time (TAI) with leap seconds added at irregular intervals to compensate for the slowing of Earth's rotation. UTC is not adjusted for daylight saving time. It is effectively a successor to Greenwich Mean Time (GMT).

Overview Map: Path of Totality – Russia, Greenland, Iceland, and Spain

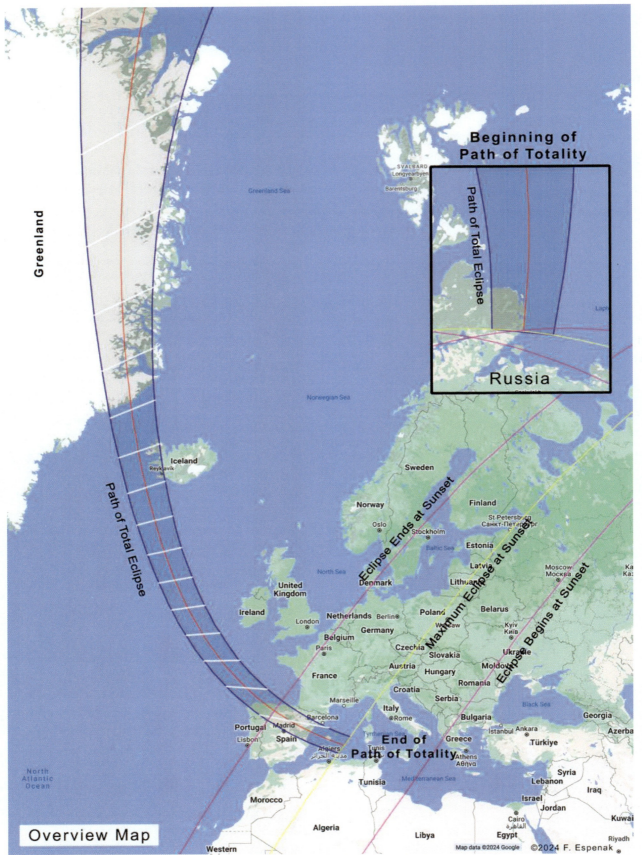

Figure 3 – Overview Map: Path of Totality – Russia, Greenland, Iceland, and Spain

4 – Path of Totality

The path of totality begins at 17:00 UTC in Russia's Taymyr Peninsula. Along this frigid and uninhabited Arctic coastline, the central line duration of totality is 1 min 34 sec (1 minute 34 seconds) at sunrise. The Moon's umbral shadow races north with an initial velocity of over 36 km/s (81,000 mph).

By 17:06 UTC, the shadow has traveled 1560 km (970 miles) as it passes within 100 km (62 miles) of the North Pole. The duration of totality is 1 min 54 sec and the shadow's velocity has dropped to 2.1 km/sec (4700 mph).

The trajectory of the track takes it south to encounter the northeastern coastline of Greenland at 17:15 UTC. The central line duration of totality is 2 min 6 sec and the Sun stands 19° above the horizon. The path width is 274 km (170 miles) as the shadow pursues its southern track with a ground velocity of 0.70 km/s (1566 mi/her).

The mighty fjords – Independence, Hagans and Denmark – all lie within the path as it leaves the coast and travels inland and up onto Greenland's icecap. By 17:30 UTC, the umbra is returning to Greenland's eastern coastline along Scoresby Sound, the longest fjord system on Earth. The duration of totality is 2 min 15 sec and the Sun's altitude is 24°.

Leaving Greenland, the eclipse track crosses the Denmark Strait where the instant of Greatest Eclipse occurs at 17:45:57 UTC. The duration of totality is 2 min 18 sec and the Sun's altitude is 25.8°. At the same time, the eastern edge of the eclipse track is racing across the rugged western coastline of Iceland. Parts of Vestfiroir, Vesturland, and Suournes all extend into the eclipse path although none of them reach as far as the central line. The capital city Reykjavik is within the path where totality occurs at 17:48 UTC and lasts 59 seconds with the Sun 24° above the southwest horizon.

The eclipse track crosses the North Atlantic where it passes 500 km west of Ireland at about 18:10 UTC.

Figure 4 – The path of totality crosses the Iberian Peninsula and ends in the Mediterranean Sea.

By 18:27 UTC, the path of totality reaches the north coast of the Iberian Peninsula (Figure 4). The Spanish city Oviedo lies just north of the central line. Mid-totality here is at 18:28 UTC (8:28 pm CEST) and lasts for 1 min 48 sec while the late afternoon Sun stands just 10° above the horizon.

As the shadow's trajectory takes it across the Iberian Peninsula, it nicks the northeast corner of Portugal. Only about 32 square km (12 square miles) of rural Portuguese real estate lies within the path. The only municipality of any size is Braganca where a partial eclipse of magnitude 0.998 will be seen.

*Figure 5 – The southern limit of the path of totality passes just north of Madrid.
If Madrileños travel a short distance north, they can easily reach the path and experience totality.*

Spain's capital city Madrid also lies just outside the totality track's southern limit. Maximum eclipse from the city center occurs at 08:32:26 pm CEST where a partial eclipse of magnitude 0.999 will be visible. However, metropolitan communities to the north like Alcobendas are located just inside the path and receive 20 seconds of totality (Figure 5).

As the Moon's shadow traverses the Iberian Peninsula, the duration of totality and the Sun's altitude gradually decrease. By the time it reaches the Mediterranean coast (18:31 GMT or 8:31 pm CEST), the central line duration is 1 min 39 sec but the Sun's altitude is only 4.5°. Valencia, Spain's largest city in the path, is located 114 km (70 miles) south of the center. Its 843,000 inhabitants will be treated to totality lasting about 60 seconds.

Continuing across the western Mediterranean, the path encounters its last landfall with Spain's Balearic Islands of Ibiza, Mallorca, and Menorca. Here the Sun's altitude is only about 2° above the Sea.
The shadow reaches the end of its track at 18:32 UTC as it returns to space. Over a period of 1 hour 32 minutes, the umbra travels along a track length of 8260 km (5130 miles) that covers an area just 0.47% of Earth's surface. The coordinates of the central line and additional information about the path are presented in Table 1.

Table 1 – Path of Totality

UTC	Latitude	Longitude	Diameter Ratio	Sun Alt	Sun Azm	Path Width km	Major Axis km	Minor Axis km	Duration Totality
17:00.1	75°04.7'N	113°26.7'E	1.031	0.0°	7.0°	268.7	—	105.1	01m34.3s
17:05	86°22.2'N	096°10.5'E	1.035	11.2°	351.3°	274.3	601.5	116.4	01m52.7s
17:10	86°51.9'N	001°12.3'W	1.036	15.6°	255.9°	275.0	450.1	120.6	02m00.4s
17:15	83°14.3'N	022°55.0'W	1.037	18.6°	236.3°	276.0	387.1	123.5	02m05.8s
17:20	79°48.2'N	026°58.9'W	1.037	20.9°	234.4°	277.5	351.9	125.6	02m10.0s
17:25	76°38.9'N	027°54.6'W	1.038	22.7°	235.8°	279.6	330.0	127.2	02m13.1s
17:30	73°42.6'N	027°48.5'W	1.038	24.0°	238.2°	282.2	315.6	128.3	02m15.4s
17:35	70°55.8'N	027°14.8'W	1.038	24.9°	241.2°	285.4	306.4	129.0	02m17.0s
17:40	68°16.2'N	026°25.9'W	1.039	25.5°	244.4°	289.0	300.9	129.5	02m17.9s
17:45	65°42.0'N	025°26.7'W	1.039	25.7°	247.7°	293.1	298.5	129.6	02m18.2s
17:50	63°11.7'N	024°18.7'W	1.039	25.7°	251.2°	297.6	299.0	129.4	02m17.9s
17:55	60°44.0'N	023°02.2'W	1.038	25.3°	254.7°	302.5	302.5	128.9	02m16.9s
18:00	58°17.6'N	021°36.1'W	1.038	24.5°	258.3°	307.4	309.3	128.1	02m15.3s
18:05	55°51.3'N	019°58.3'W	1.038	23.4°	261.9°	312.2	320.4	127.0	02m13.0s
18:10	53°23.6'N	018°05.4'W	1.037	21.8°	265.5°	316.4	337.6	125.4	02m10.0s
18:15	50°52.4'N	015°51.4'W	1.037	19.8°	269.3°	319.1	364.6	123.4	02m06.2s
18:20	48°14.2'N	013°05.5'W	1.036	17.2°	273.2°	319.0	410.2	120.7	02m01.2s
18:25	45°22.0'N	009°23.5'W	1.035	13.5°	277.5°	313.2	502.5	117.0	01m54.7s
18:30	41°51.4'N	003°17.2'W	1.033	7.7°	283.0°	294.3	839.9	111.0	01m44.7s
18:32.2	38°40.7'N	005°24.1'E	1.031	0.0°	289.1°	261.1	—	103.2	01m32.7s

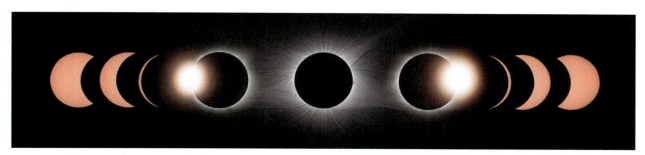

Photo 3 – Time sequence of the Total Solar Eclipse of 2017 August 21.
This sequence shows the partial phases, the diamond ring effect, and the solar corona during totality.
© 2017 F. Espenak, www.MrEclipse.com

5 – Local Circumstances

Local circumstances for the **Total Eclipse** for a number of cities are found in Table 2 (Iceland) and Table 3 (Spain). All times are given in local time – *Greenwich Mean Time (GMT)* in Iceland, and *Central European Summer Time (CEST)* in Spain. The *Duration Totality* is period between the beginning and end of totality. These times include the effects on the Moon's limb profile. The Sun's altitude is given at the instant of mid-totality.

Local circumstances for the **Partial Eclipse** for many cities are found in Table 4 (Europe), Table 5 (Africa), Table 6 (USA), and Table 7 (Canada). All times are given in local time and include *Daylight Saving Time* (aka *Summer Time* in Europe). The Sun's altitude is given at the instant of maximum eclipse. *Eclipse Magnitude* is given as the percentage of the Sun's *diameter* covered by the Moon at maximum eclipse.

The 2026 Total Solar Eclipse Circumstances Calculator is an interactive web page that can quickly calculate the local circumstances for the solar eclipse from any geographic location not included in Tables 2 through 7. The URL address for the calculator is:

https://www.EclipseWise.com/solar/SEcirc/2001-2100/SE2026Aug12Tcirc.html

Table 2 - Local Circumstances for the Total Eclipse in Iceland

City	Partial Eclipse Begins	Total Eclipse Begins	Total Eclipse Ends	Partial Eclipse Ends	Duration Totality	Sun Altitude
Akranes	04:46:56 pm	05:47:59 pm	05:49:01 pm	06:47:22 pm	01m02s	25°
Flatey	04:44:39 pm	05:45:29 pm	05:47:10 pm	06:45:25 pm	01m41s	25°
Flateyri	04:43:14 pm	05:44:00 pm	05:45:52 pm	06:44:07 pm	01m53s	25°
Grindavik	04:47:31 pm	05:48:24 pm	05:50:01 pm	06:48:10 pm	01m37s	25°
Isafjordur	04:43:11 pm	05:44:00 pm	05:45:33 pm	06:43:52 pm	01m33s	25°
Keflavik	04:47:09 pm	05:48:02 pm	05:49:42 pm	06:47:51 pm	01m40s	25°
Olafsvik	04:45:07 pm	05:45:56 pm	05:47:59 pm	06:46:09 pm	02m03s	25°
Reykjavik	04:47:17 pm	05:48:24 pm	05:49:14 pm	06:47:40 pm	00m50s	24°
Stykkisholmur	04:45:10 pm	05:46:03 pm	05:47:34 pm	06:45:50 pm	01m31s	25°

Table 3 - Local Circumstances for the Total Eclipse in Spain

City	Partial Eclipse Begins	Total Eclipse Begins	Total Eclipse Ends	Partial Eclipse Ends	Duration Totality	Sun Altitude
Alcala de Henares	07:36:40 pm	08:31:54 pm	08:32:32 pm	-	00m38s	7°
Alcobendas	07:36:34 pm	08:32:00 pm	08:32:23 pm	-	00m23s	7°
Aviles	07:31:01 pm	08:26:49 pm	08:28:36 pm	09:20:51 pm	01m47s	10°
Benavente	07:33:49 pm	08:29:33 pm	08:30:55 pm	09:23:04 pm	01m22s	9°
Bilbao	07:31:50 pm	08:27:23 pm	08:27:58 pm	-	00m35s	8°
Burgos	07:33:23 pm	08:28:25 pm	08:30:09 pm	-	01m44s	8°
Castellon de la Plana	07:37:33 pm	08:31:20 pm	08:32:54 pm	-	01m34s	5°
Cuenca	07:37:26 pm	08:32:08 pm	08:33:00 pm	-	00m52s	6°
El Ferrol del Caudillo	07:30:47 pm	08:27:23 pm	08:28:52 pm	09:21:47 pm	01m29s	12°
Gijon	07:31:04 pm	08:26:49 pm	08:28:34 pm	09:20:47 pm	01m45s	10°
Guadalajara	07:36:25 pm	08:31:22 pm	08:32:29 pm	-	01m06s	7°
Ibiza	07:39:18 pm	08:32:47 pm	08:33:50 pm	-	01m03s	3°
La Coruna	07:30:58 pm	08:27:42 pm	08:28:59 pm	09:22:01 pm	01m17s	12°
Leon	07:32:46 pm	08:28:21 pm	08:30:06 pm	09:22:08 pm	01m45s	10°
Lerida	07:34:46 pm	08:29:11 pm	08:29:39 pm	-	00m28s	5°
Logrono	07:33:15 pm	08:28:10 pm	08:29:31 pm	-	01m21s	8°
Lugo	07:31:47 pm	08:28:10 pm	08:29:33 pm	09:22:17 pm	01m23s	11°
Menorca	07:37:15 pm	08:30:13 pm	08:31:22 pm	-	01m08s	2°
Oviedo	07:31:21 pm	08:27:06 pm	08:28:54 pm	09:21:05 pm	01m48s	10°
Palencia	07:33:54 pm	08:29:09 pm	08:30:51 pm	-	01m42s	9°
Palma (de Mallorca)	07:38:05 pm	08:31:06 pm	08:32:43 pm	-	01m36s	3°
Portugalete	07:31:43 pm	08:27:21 pm	08:27:48 pm	-	00m26s	8°
Reus	07:35:32 pm	08:29:28 pm	08:30:30 pm	-	01m02s	4°
San Sebastian de Reyes	07:36:33 pm	08:31:56 pm	08:32:24 pm	-	00m28s	7°
Santander	07:31:23 pm	08:26:58 pm	08:28:00 pm	09:20:08 pm	01m02s	9°
San Vicente Baracaldo	07:31:45 pm	08:27:23 pm	08:27:49 pm	-	00m26s	8°
Segovia	07:35:49 pm	08:31:10 pm	08:32:06 pm	-	00m57s	8°
Soria	07:34:28 pm	08:29:07 pm	08:30:49 pm	-	01m42s	7°
Tarragona	07:35:35 pm	08:29:30 pm	08:30:30 pm	-	01m00s	4°
Torrejon de Ardoz	07:36:44 pm	08:32:09 pm	08:32:26 pm	-	00m17s	7°
Torrente	07:38:30 pm	08:32:39 pm	08:33:32 pm	-	00m52s	5°
Valencia	07:38:26 pm	08:32:31 pm	08:33:30 pm	-	01m00s	5°
Valladolid	07:34:32 pm	08:29:54 pm	08:31:21 pm	-	01m27s	9°
Vitoria	07:32:34 pm	08:27:44 pm	08:28:47 pm	-	01m03s	8°
Zamora	07:34:43 pm	08:30:56 pm	08:31:11 pm	09:23:49 pm	00m15s	9°
Zaragoza	07:34:44 pm	08:29:05 pm	08:30:30 pm	-	01m25s	6°

Table 4 - Local Circumstances for the Partial Eclipse in Europe

Country	City	Partial Eclipse Begins	Maximum Eclipse	Partial Eclipse Ends	Eclipse Magnitude	Sun Altitude
Austria	Vienna	07:22:05 pm	08:13:52 pm	-	89%	0°
Belarus	Minsk	08:10:16 pm	-	-	76%	0°
Belgium	Brussels	07:18:50 pm	08:13:39 pm	09:05:37 pm	91%	8°
Croatia	Zagreb	07:25:57 pm	-	-	79%	0°
Czech Republic	Prague	07:19:25 pm	08:11:49 pm	-	88%	2°
Denmark	Copenhagen	07:10:11 pm	08:03:39 pm	-	86%	6°
Finland	Helsinki	08:00:50 pm	08:52:42 pm	-	83%	3°
France	Bordeaux	07:29:12 pm	08:24:35 pm	-	97%	7°
	Lille	07:19:09 pm	08:14:15 pm	09:06:28 pm	92%	8°
	Lyon	07:27:37 pm	08:21:43 pm	-	94%	5°
	Marseille	07:31:41 pm	08:25:21 pm	-	96%	3°
	Paris	07:22:14 pm	08:17:20 pm	09:09:27 pm	93%	8°
Germany	Berlin	07:15:32 pm	08:08:28 pm	-	87%	4°
	Cologne	07:18:39 pm	08:12:50 pm	-	90%	6°
	Dusseldorf	07:18:12 pm	08:12:27 pm	-	90%	6°
	Essen	07:17:44 pm	08:11:58 pm	-	90%	6°
	Frankfurt	07:20:00 pm	08:13:41 pm	-	90%	5°
	Hamburg	07:14:01 pm	08:07:48 pm	-	88%	6°
	Hannover	07:16:01 pm	08:09:43 pm	-	88%	5°
	Mannheim	07:21:05 pm	08:14:45 pm	-	90%	4°
	Munich	07:23:02 pm	08:15:50 pm	-	90%	2°
	Stuttgart	07:22:14 pm	08:15:39 pm	-	91%	4°
Hungary	Budapest	07:22:32 pm	-	-	78%	0°
Ireland	Dublin	06:12:58 pm	07:10:44 pm	08:05:20 pm	95%	15°
Italy	Genoa	07:29:27 pm	08:22:25 pm	-	94%	1°
	Milan	07:27:43 pm	08:20:44 pm	-	93%	2°
	Naples	07:33:57 pm	-	-	62%	0°
	Rome	07:32:48 pm	-	-	85%	0°
	Turin	07:28:34 pm	08:21:54 pm	-	94%	3°
Latvia	Riga	09:06:20 pm	09:57:50 pm	-	84%	1°
Lithuania	Vilnius	09:09:41 pm	-	-	82%	0°
Macedonia	Skopje	07:29:59 pm	-	-	26%	0°
Netherlands	Amsterdam	07:16:08 pm	08:10:58 pm	09:03:03 pm	90%	8°
Norway	Oslo	07:02:42 pm	07:57:04 pm	08:49:13 pm	86%	9°
Poland	Krakow	07:18:21 pm	-	-	87%	0°
	Warsaw	07:14:37 pm	08:06:00 pm	-	86%	0°
Portugal	Lisbon	06:39:21 pm	07:36:10 pm	08:29:09 pm	95%	10°
	Porto	06:34:56 pm	07:32:02 pm	08:25:22 pm	98%	11°
Romania	Bucharest	08:24:41 pm	-	-	5%	0°
Russia	Moscow	08:03:55 pm	-	-	25%	0°
	St. Petersburg	08:00:02 pm	08:51:04 pm	-	83%	1°
Serbia & Mont.	Beograd	07:26:09 pm	-	-	50%	0°
Spain	Barcelona	07:35:06 pm	08:29:19 pm	-	99%	4°
	Madrid	07:36:49 pm	08:32:26 pm	-	99%	7°
	Malaga	07:43:12 pm	08:38:21 pm	-	95%	6°
	Sevilla	07:42:00 pm	08:37:42 pm	-	95%	7°
Sweden	Stockholm	07:03:19 pm	07:56:13 pm	08:47:05 pm	84%	5°
Switzerland	Zurich	07:24:37 pm	08:18:02 pm	-	92%	3°
UK	Bristol	06:17:08 pm	07:13:47 pm	08:07:19 pm	94%	12°
	Birmingham	06:15:20 pm	07:11:52 pm	08:05:22 pm	93%	12°
	Glasgow	06:08:38 pm	07:06:04 pm	08:00:32 pm	92%	14°
	Leeds	06:12:55 pm	07:09:30 pm	08:03:08 pm	92%	12°
	Liverpool	06:13:28 pm	07:10:22 pm	08:04:14 pm	93%	13°
	London	06:17:21 pm	07:13:22 pm	08:06:23 pm	93%	10°
	Manchester	06:13:25 pm	07:10:10 pm	08:03:53 pm	92%	12°
	Newcastle u. Tyne	06:10:48 pm	07:07:29 pm	08:01:16 pm	92%	13°
Ukraine	Kiev	09:14:23 pm	-	-	14%	0°

Table 5 - Local Circumstances for the Partial Eclipse in Africa

Country	City	Partial Eclipse Begins	Maximum Eclipse	Partial Eclipse Ends	Eclipse Magnitude	Sun Altitude
Malta	Valletta	07:41:12 pm	-	-	31%	0°
Algeria	Algiers	06:42:36 pm	07:35:51 pm	-	98%	1°
Libya	Tripoli	06:46:00 pm	-	-	19%	0°
Morocco	Casablanca	06:48:31 pm	07:43:48 pm	-	89%	7°
	Fes	06:47:45 pm	07:42:27 pm	-	91%	5°
	Rabat	06:47:51 pm	07:43:03 pm	-	90%	6°
Senegal	Dakar	06:25:16 pm	07:12:13 pm	-	47%	4°
Sierra Leone	Freetown	06:36:55 pm	-	-	34%	0°
Tunisia	Tunis	06:41:10 pm	-	-	67%	0°

Table 6 - Local Circumstances for the Partial Eclipse in the USA

Country	City	Partial Eclipse Begins	Maximum Eclipse	Partial Eclipse Ends	Eclipse Magnitude	Sun Altitude
Alaska	Anchorage	06:36:13 am	07:22:02 am	08:09:50 am	39%	15°
Connecticut	Bridgeport	01:05:50 pm	01:54:26 pm	02:41:13 pm	21%	61°
	Hartford	01:03:38 pm	01:54:01 pm	02:42:29 pm	23%	60°
	New Haven	01:05:24 pm	01:54:37 pm	02:41:58 pm	22%	61°
Dist. of Columbia	Washington	01:17:50 pm	01:53:05 pm	02:27:28 pm	10%	64°
Illinois	Chicago	12:16:43 pm	12:28:23 pm	12:40:01 pm	1%	62°
Indiana	Fort Wayne	01:16:39 pm	01:34:29 pm	01:52:13 pm	2%	64°
	South Bend	01:13:31 pm	01:31:21 pm	01:49:07 pm	2%	63°
Maryland	Baltimore	01:15:30 pm	01:52:57 pm	02:29:23 pm	11%	64°
Massachusetts	Boston	01:01:49 pm	01:55:08 pm	02:46:13 pm	27%	59°
	Springfield	01:02:25 pm	01:53:31 pm	02:42:38 pm	24%	60°
	Worcester	01:01:58 pm	01:54:18 pm	02:44:32 pm	25%	59°
Michigan	Detroit	01:03:19 pm	01:36:10 pm	02:08:35 pm	9%	62°
	Flint	12:59:43 pm	01:33:53 pm	02:07:36 pm	9%	62°
	Grand Rapids	01:01:36 pm	01:30:22 pm	01:58:54 pm	6%	62°
	Lansing	01:02:04 pm	01:32:47 pm	02:03:11 pm	7%	62°
	Saginaw	12:57:31 pm	01:32:45 pm	02:07:33 pm	10%	61°
Minnesota	Minneapolis	11:55:19 am	12:13:23 pm	12:31:26 pm	3%	57°
	St. Paul	11:55:24 am	12:13:42 pm	12:31:59 pm	3%	57°
New Jersey	Newark	01:07:43 pm	01:53:53 pm	02:38:27 pm	18%	62°
New York	Albany	01:00:10 pm	01:50:44 pm	02:39:29 pm	23%	60°
	Buffalo	12:58:54 pm	01:42:19 pm	02:24:41 pm	16%	62°
	New York	01:07:46 pm	01:54:10 pm	02:38:56 pm	19%	62°
	Rochester	12:57:41 pm	01:43:50 pm	02:28:45 pm	18%	61°
	Syracuse	12:58:16 pm	01:46:21 pm	02:32:58 pm	20%	61°
Ohio	Akron	01:09:35 pm	01:41:07 pm	02:12:09 pm	8%	64°
	Canton	01:11:15 pm	01:41:52 pm	02:12:02 pm	7%	64°
	Cleveland	01:07:12 pm	01:40:02 pm	02:12:21 pm	8%	63°
	Columbus	01:21:02 pm	01:40:29 pm	01:59:47 pm	3%	65°
	Toledo	01:08:16 pm	01:36:27 pm	02:04:19 pm	6%	63°
	Youngstown	01:08:39 pm	01:42:36 pm	02:15:57 pm	9%	64°
Pennsylvania	Allentown	01:08:27 pm	01:52:05 pm	02:34:21 pm	16%	62°
	Harrisburg	01:10:37 pm	01:50:33 pm	02:29:23 pm	13%	63°
	Philadelphia	01:11:24 pm	01:53:55 pm	02:35:05 pm	15%	63°
	Pittsburgh	01:12:02 pm	01:44:58 pm	02:17:17 pm	9%	64°
	Scranton	01:04:58 pm	01:50:13 pm	02:34:04 pm	17%	62°
Rhode Island	Providence	01:03:40 pm	01:55:45 pm	02:45:42 pm	25%	60°
Virginia	Newport News	01:28:55 pm	01:58:01 pm	02:26:29 pm	7%	65°
	N. Springfield	01:18:45 pm	01:52:56 pm	02:26:19 pm	9%	65°
	Richmond	01:27:01 pm	01:55:06 pm	02:22:39 pm	6%	66°
	Virginia Beach	01:29:00 pm	01:59:00 pm	02:28:19 pm	7%	65°
Wisconsin	Madison	12:07:36 pm	12:23:11 pm	12:38:44 pm	2%	61°
	Milwaukee	12:04:15 pm	12:26:04 pm	12:47:48 pm	4%	61°

Table 7 - Local Circumstances for the Partial Eclipse in Canada

Country	City	Partial Eclipse Begins	Maximum Eclipse	Partial Eclipse Ends	Eclipse Magnitude	Sun Altitude
Alberta	Calgary	10:11:06 am	10:36:55 am	11:03:13 am	7%	38°
	Edmonton	10:02:14 am	10:38:37 am	11:16:01 am	14%	38°
	Lethbridge	10:20:32 am	10:38:12 am	10:56:02 am	3%	40°
British Co.	Kamloops	09:09:03 am	09:29:10 am	09:49:34 am	4%	34°
	Kelowna	09:15:50 am	09:29:42 am	09:43:38 am	2%	35°
	Prince George	08:54:39 am	09:28:48 am	10:04:03 am	14%	31°
Manitoba	Brandon	11:24:10 am	11:57:51 am	12:31:57 pm	10%	49°
	Churchill	11:05:52 am	12:02:10 pm	12:59:05 pm	39%	44°
	Winnipeg	11:25:06 am	12:02:21 pm	12:40:00 pm	13%	50°
N. Brunswick	Moncton	01:53:44 pm	02:55:49 pm	03:54:46 pm	43%	53°
	Saint John	01:55:11 pm	02:55:56 pm	03:53:42 pm	40%	54°
Newfound.	Gander	02:24:19 pm	03:30:38 pm	04:33:01 pm	61%	46°
	Saint John's	02:28:47 pm	03:34:59 pm	04:37:01 pm	62%	45°
Northwest T.	Aklavik	09:42:14 am	10:35:41 am	11:31:02 am	54%	22°
	Inuvik	09:42:45 am	10:36:32 am	11:32:12 am	54%	23°
	Yellowknife	09:49:41 am	10:42:18 am	11:36:35 am	40%	33°
Nova Scotia	Dartmouth	01:58:19 pm	03:00:02 pm	03:58:25 pm	42%	53°
	Halifax	01:58:20 pm	03:00:02 pm	03:58:23 pm	42%	53°
	Sydney	01:56:41 pm	03:00:47 pm	04:01:13 pm	50%	50°
Nunavut	Alert	12:09:51 pm	01:09:32 pm	02:08:56 pm	94%	22°
	Iqaluit	12:18:53 pm	01:24:10 pm	02:28:07 pm	69%	40°
	Resolute	10:59:20 am	12:00:14 pm	01:01:37 pm	74%	29°
	Grise Fiord	12:04:21 pm	01:05:50 pm	02:07:21 pm	81%	28°
Ontario	Brampton	12:55:28 pm	01:39:26 pm	02:22:27 pm	16%	61°
	Burlington	12:57:04 pm	01:40:03 pm	02:22:06 pm	16%	61°
	Etobicoke	12:55:37 pm	01:39:49 pm	02:23:03 pm	17%	61°
	Hamilton	12:57:22 pm	01:40:04 pm	02:21:50 pm	15%	61°
	Kingston	12:53:43 pm	01:43:38 pm	02:32:05 pm	22%	60°
	Kitchener	12:56:31 pm	01:38:39 pm	02:19:57 pm	15%	61°
	London	12:58:52 pm	01:38:12 pm	02:16:50 pm	13%	62°
	Markham	12:54:42 pm	01:39:56 pm	02:24:07 pm	18%	61°
	Mississauga	12:55:54 pm	01:39:51 pm	02:22:51 pm	16%	61°
	North York	12:55:07 pm	01:39:52 pm	02:23:36 pm	17%	61°
	Oshawa	12:54:36 pm	01:40:33 pm	02:25:24 pm	18%	61°
	Ottawa	12:49:51 pm	01:42:43 pm	02:33:58 pm	26%	59°
	St. Catharines	12:57:42 pm	01:41:12 pm	02:23:43 pm	16%	61°
	Scarborough	12:55:16 pm	01:40:10 pm	02:24:02 pm	17%	61°
	Sudbury	12:44:07 pm	01:32:45 pm	02:20:29 pm	21%	58°
	Thunder Bay	12:34:12 pm	01:16:43 pm	01:59:09 pm	16%	55°
	Toronto	12:55:36 pm	01:40:07 pm	02:23:38 pm	17%	61°
	Waterloo	12:56:28 pm	01:38:34 pm	02:19:51 pm	15%	61°
	Windsor	01:03:29 pm	01:36:17 pm	02:08:38 pm	8%	63°
	York	12:55:28 pm	01:39:54 pm	02:23:20 pm	17%	61°
P.E.I.	Charlottetown	01:54:29 pm	02:57:26 pm	03:57:05 pm	46%	52°
Quebec	Chicoutimi	12:43:35 pm	01:43:47 pm	02:41:46 pm	39%	55°
	Laval	12:50:09 pm	01:45:14 pm	02:38:25 pm	29%	58°
	Longueuil	12:50:25 pm	01:45:41 pm	02:39:00 pm	29%	58°
	Montreal	12:50:26 pm	01:45:37 pm	02:38:52 pm	29%	58°
	Quebec	12:47:49 pm	01:46:27 pm	02:42:50 pm	35%	56°
	Sherbrooke	12:51:36 pm	01:48:09 pm	02:42:32 pm	31%	57°
	Trois Rivières	12:48:25 pm	01:45:31 pm	02:40:33 pm	32%	57°
Saskatchewan	Moose Jaw	10:19:11 am	10:48:56 am	11:19:11 am	8%	45°
	Regina	10:19:26 am	10:50:17 am	11:21:37 am	9%	45°
	Saskatoon	10:11:22 am	10:47:15 am	11:23:53 am	13%	43°
Yukon	Dawson	08:39:14 am	09:28:48 am	10:20:25 am	44%	21°
	Whitehorse	08:40:23 am	09:26:38 am	10:14:50 am	34%	23°

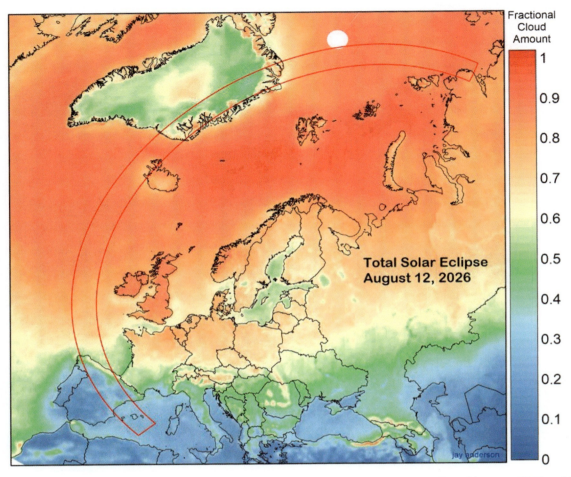

Figure 6 – Polar-satellite-derived cloud-cover estimates for August based on observations collected between 2001 and 2021. Data: CM SAF/EUMETSAT.

6 – Weather Prospects for the 2026 Total Solar Eclipse

The following is an abridged version a report by Jay Anderson.

RUSSIA/SIBERIA

The track of the 2026 total eclipse begins at sunrise over the polar permafrost climate of Russia's Taymyr Peninsula and the nearby Laptev Sea. This coast has a mean August cloud amount of around 75%. The weather station in nearby Dikson (80% partial eclipse) show an average August sunshine of 21%. Even less enticing, the data reports 21 days of the month with rain or snow and an average high temperature of 8°C. All-in-all, an unforgiving location to view an eclipse.

GREENLAND

After crossing the Arctic Ocean, the path passes 100 km south of the North Pole. The route is plagued by heavy cloudiness, as shown by the orange tones in Figure 6. The High Arctic climatology in summer is characterized by extensive and persistent cloudiness.

Although weather prospects improve in Greenland, the best weather is found high up on the icecap. However, as the cold, dry air spills off the icecap into the fjords, the air is warmed and further dried by its descent. The deepest of the inlets, Scoresby Sund, slice right to the central line of the eclipse track and even a little beyond, sampling the clear, dry air that descends from the frozen heights.

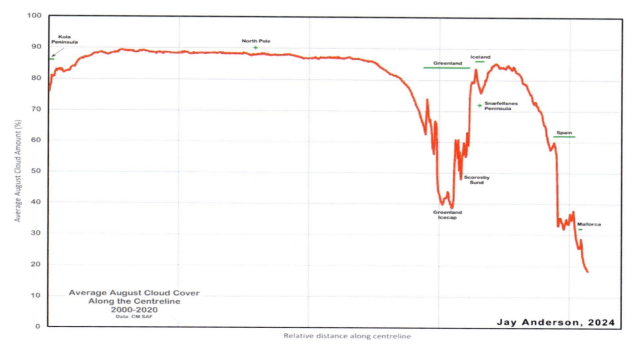

Graph 1 – Cloud amount in percent along the central line. These data were extracted from the same source as Figure 6. Important geographical regions are labelled along with their extent where appropriate. Cloud amounts for the North Pole and the Snæfellsnes Peninsula, which are not on the central line, are shown as small crosses at their individual cloud values.

Satellite measurements of August cloud cover along the Greenland coast show that the deep reaches of Scoresby Sund have a mean cloudiness just under 50%, as shown in Graph 1. Ittoqqortoormiit, a village at the entrance to Scoresby Sund, records an average August sunshine at 40%.

Over the 21-years of Aqua satellite images (around 1:30 p.m. local time) they show the eclipse would have been easily visible in clear skies 61% of the time from Scoresby Sund and visible through thin clouds on another 19%. That's an impressive success rate of 80%. While 21 days is a small sample, it demonstrates that Greenland's interior inlets might be very good eclipse-watching sites.

ICELAND

Iceland is a destination of striking beauty with volcanoes, glaciers, and aurorae. Its terrain is complex, with strong meteorological influences from both elevation and proximity to water. The island is actually sunnier than the nearby waters of the Iceland Sea. Cloud cover averages about 70 to 80% over land compared to the 80 to 90% offshore. Cloud cover over the land tends to be convective, broken up by the flow over the terrain and by solar heating. Over-water cloud tends to come in large broken to overcast sheets.

Wind direction and speed play key roles in Icelandic weather forecasting. Air piles cloud up on windward sides of the island and opens skies on the leeward side. Classically, one or more of the several peninsulas that extend into the eclipse zone will have dense cloud on their upwind side and nearly clear skies on their lee.

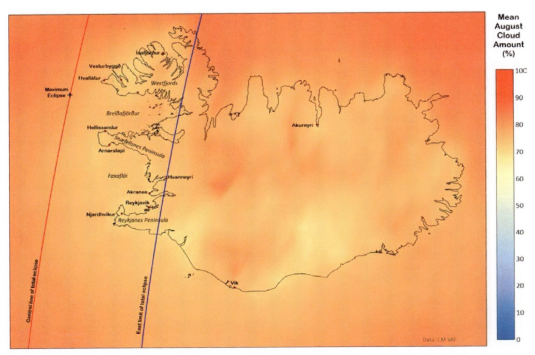

Figure 7 – Average August cloud cover over western Iceland. Data: CM SAF/EUMETSAT

The western peninsulas — Westfords, Snaefellsness, and Reykjanes — are sensitive to convective clouds, build along the peninsula mid-lines on sunny days, leaving the coast in clear skies. The eclipse choices include of watching from Reykjavik, heading northwest to get as deep as possible into the shadow, or, best of all, following the forecasts the days before the eclipse to find the most promising spot. The Icelandic Met Office displays a cloud-cover forecast map on its web site which could prove invaluable leading up to eclipse day (***https://en.vedur.is/weather/forecasts/cloudcover/***).

Sunshine measurements from Reykjavik are 32% in August. This fits well with the mean cloudiness of 71% as measured from satellite.

SPAIN AND THE BALEARIC ISLANDS

Spain's summer climate is one where sunshine dominates, with August's possible sunshine generally lying in the 70% range along the eclipse track. It's not an evenly spread cloudiness, as certain parts of the country are more sensitive to the formation of cloud than others. There is value in paying attention to the nuances of Spain's climatology and topography together.

Eclipse viewing is complicated by the low Sun elevation, only 10° on the north coast and a mere 4° at the Balearic Sea during totality. A clear view of the horizon in the sun's direction is critical. Care must be taken to avoid hills and other obstacles along the horizon.

The Iberian Peninsula along the eclipse path has three mountain ranges, two elevated central plateaus, a river valley, and two coastal plains. In the north, the Cordillera Cantabrica blocks moisture from the Bay of Biscay, holding it against the coastal lowlands and preventing its inland. There are gaps in the barrier, primarily on the north side of the central line, between Bilbao and Burgos, but even there, the moisture invasion is usually slowed by the 1000-m heights northeast of Burgos.

In central Spain, the eclipse track moves onto a more complicated landscape. The north side of the path features lowlands of the Ebro River Basin. On the south, the Meseta Central, a 600-700 meter plateau is split a north/south by the Sierra de Guadarrama, a range of mountains north of Madrid.

The Sistema Iberico begins southeast of Burgos and arcs southward as it approaches the Mediterranean coast. It is a haphazard collection of smaller ranges and massifs. Peaks in the Iberico are modest, reaching ~2300 m.

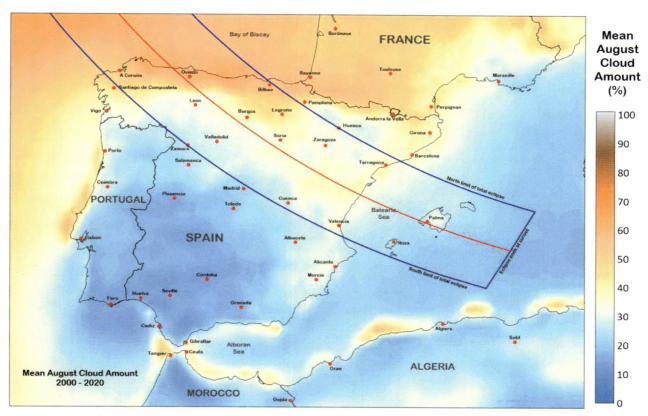

Figure 8 – Average August cloud cover as measured from polar-orbiting satellites between 2002 and 2020. Data: CM SAF/EUMETSAT.

CLOUDS AND WEATHER IN SPAIN

The varying topography has an impact on the cloud climatology of the Iberian Peninsula. In Figure 8, we see the broad outlines of the terrain mirrored in the cloud patterns. Higher-cloud yellow tones largely lie atop the elevated terrain, while blue shades spread across the lower elevations of the Meseta basins and the Ebro Valley. The differences in cloud amount are relatively small once the track has crossed the Cordillera Cantabrica, especially along the central line, which dodges between the good- and poor-weather regions but always stays around 35%.

In Figure 8, we see that the north coast along the Bay of Biscay is an uninviting destination for eclipse viewing. Satellite-measured cloud amounts are close to 60%.

The Cantabricas are highest on the south side of the track and across the eclipse midline, but peter out toward the north limit, allowing Atlantic cloud and weather to move farther inland. Graph 2 shows an abrupt transition from cloudy to sunny along the central line.

Cloud over the north coast comes from a variety of sources. Mostly, it's just a pile-up of stratus and stratocumulus clouds driven ashore onto the mountain slopes by northerly winds. In addition to the cloud, exposure to a cold ocean brings temperatures that average several degrees lower than the rest of the Iberian eclipse track and rain comes two to four times more often in the month.

Once over the Cordillera Catabrica, the eclipse track adopts a Jekyll-and-Hyde personality. Wherever the track crosses forested terrain, cloudiness increases; on the flatter and lower parts, it declines. In Figure 8, the cloud over the Sistema Iberico is visible as a pale yellow band from Logroño to the coast north of Valencia. About half-way along, a pale orange finger of higher cloud extends southwestward from Soria almost to Pladencia, passing north of Madrid.

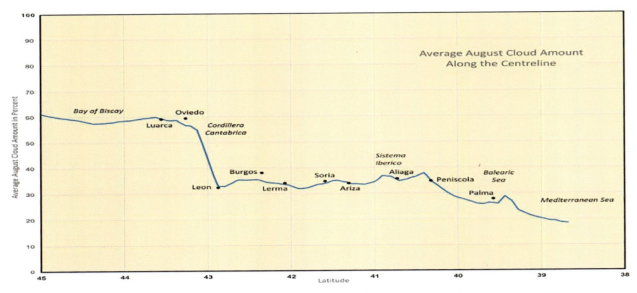

Graph 2 – A cloud-cover along the central line of the path of totality through Spain. Data: NASA.

Leon, Burgos, and Valladolid are three prominent destinations in the north basin of the Meseta Central and show average August sunshine amounts of 68 to 78%. Two decades of satellite data suggests an easily visible eclipse 81% of the time. In the south basin, Madrid, Cuenca, and Guadalajara lie in the sunniest zone.

The various mountain ranges of the Sistema Iberico offer spectacular vistas toward the setting Sun, but the cloud cover map in Figure 8 suggests caution because of the low solar altitude. Satellite measurements of cloud amount ranges from 35 to 45%. Measurements of August sunshine show an average of 67% at Logroño and 73% at Soria. The difference in cloud cover between the best and poorest sites in central Spain is less than 10%.

Some of the best cloud prospects lie atop the floodplain of the Ebro River, around Huesca and Zaragoza. The central part of this region enjoys a cloud climatology that drops below 30%. Examination of daily images reveals that a site at Zaragoza would have had an easy view of the eclipse on 18 of 21 years. This region will be an attractive site for eclipse watchers although there is a penalty to pay in eclipse duration and Sun elevation.

Cloud cover on the Balearic Islands is least on the coasts, particularly on the west side, facing the setting Sun. The higher interior terrain is often dotted with cumulus clouds in the afternoon, but the cool breezes on the coast suppress them for a few kilometers inland. For such a low view toward the Sun, the skies must be very clear, a condition that is unreliable. The reward is a spectacular total eclipse only 2.5° high above the sea. Past satellite images suggest a 75% chance of success.

To read Jay Anderson's full report, see: ***https://eclipsophile.com/tse2026/***

7 – Detailed Maps of the Path of Totality

A series of 14 maps (pages 25-38) cover the entire path of totality through Russia, the Arctic, Greenland, Iceland, and Spain. The *Table of Contents* (page iii) can be used to quickly navigate to the map of interest since it lists the region featured on each map.

The map scale in Spain is approximately 1:1,400,000, which corresponds to 1 inch ≈ 22 miles (1 cm ≈ 14 km).[3] This large scale shows both major and minor roads, towns and cities, rivers, lakes, parks, national forests, wilderness areas and mountain ranges. A 50 mile (50 km) reference scale appears at the bottom of each map.

The path of totality on each map is depicted as a lightly shaded region with the northern and southern limits clearly labeled. The total eclipse can be seen only inside this path (a partial eclipse is visible outside the path). The closer one gets to the central line, the longer the total eclipse lasts. Gray lines inside the path mark the duration of the total eclipse in 30-second steps. This makes it easy to estimate the duration of totality from any location in the eclipse path.

The local time of mid-eclipse is marked by a series of white lines crossing the eclipse path every 5 minutes. Abbreviations for local times are: GMT = Greenwich Mean Time, and CEST = Central European Summer Time (= GMT + 2 hours). Eclipse circumstances on the central line are labeled with the local time of mid-eclipse, duration of totality (minutes and seconds) and altitude of the Sun.

All maps were produced using Google Maps as the underlying map with overlying eclipse graphics generated using JavaScript code. A web page is available to the user for examining any part of the 2026 eclipse path at a range of zoom magnifications. An added benefit of the web page is that it automatically calculates the local circumstances for any point the user chooses, with just the click of a mouse. For more information and to access the interactive 2026 eclipse path plotted on Google Maps, visit:

https://www.eclipsewise.com/solar/SEgmap/2001-2100/SE2026Aug12Tgmap.html

8 – EclipseWise.com Web Site

For many years the NASA Eclipse Web Site was the leading Internet resource for predictions and information on eclipses of the Sun and Moon. The webmaster of the site (Fred Espenak) has now retired it as a location for new information but it remains an archival site for eclipse predictions prior to 2015. All future predictions for upcoming eclipses will be posted on the new web site *EclipseWise.com*.

EclipseWise.com has individual web pages, maps and diagrams for every solar and lunar eclipse from 2000 BCE to 3000 CE. This covers to 11,898 solar eclipses and 12,064 lunar eclipses. Much of the design, layout and graphics were inspired by the recent publications *Thousand Year Canon of Solar Eclipses 1501 to 2500* and the *Thousand Year Canon of Lunar Eclipses 1501 to 2500*. (See: *https://www.astropixels.com/pubs*)

The graphical user interface used by *EclipseWise.com* offers an intuitive way of accessing eclipse predictions. For example, the home page presents a concise preview of all upcoming solar and lunar eclipses over several years. Each small eclipse diagram gives a quick preview of an eclipse and links to a dedicated page for that particular eclipse.

The main or top pages of EclipseWise.com are:

    Home Page (both solar and lunar eclipses): *https://www. eclipsewise.com/eclipse.html*
    Solar Eclipses Page: *https://www. eclipsewise.com/solar/solar.html*
    Lunar Eclipses Page: *https://www. eclipsewise.com/lunar/lunar.html*

---

[3] Because of the Mercator map projection, the actual scale on a given map can vary by up to 10% of this value.

EclipseWise.com and the 2026 Eclipse

EclipseWise.com has a series of pages and resources devoted to the 2026 eclipse. The main page is located at:

https://www. eclipsewise.com/solar/SEprime/2001-2100/SE2026Aug12Tprime.html

It provides links to detailed eclipse path maps, tables of eclipse path coordinates, Besselian elements, and more. The link to an interactive Google Map with the eclipse path plotted on it allows the user to zoom into an part of the path. Click on any point on the map to display the eclipse circumstances and duration of totality at that location.

Other features include information on eye safety, eclipse photography, the sky during totality and additional data tables about the eclipse path. This web site will continue to add features as the eclipse approaches.

The *2026 Total Solar Eclipse Circumstances Calculator* is an interactive web page that can quickly calculate the local circumstances for the eclipse from any geographic location. The *Calculator* is located at:

https://www.EclipseWise.com/solar/SEcirc/2001-2100/SE2026Aug12Tcirc.html

9 – Eclipse Predictions

The algorithms and software for the eclipse predictions were developed primarily from the *Explanatory Supplement to the Astronomical Ephemeris* (Her Majesty's Nautical Almanac Office, 1974) with additional algorithms from *Elements of Solar Eclipses: 1951–2200* (Meeus, 1989). The solar and lunar ephemerides were generated from the JPL DE430. All eclipse calculations were made using a value for the Moon's radius of k=0.2722810 for the path of totality. Center of mass coordinates for the Moon have been used without correction to the lunar limb profile. A value for ΔT of 69.2 seconds was used to convert the predictions from Terrestrial Time (TT) to Coordinated Universal Time (UTC).

*Photo 4 – Gazing up at the Total Solar Eclipse of 2024 April 08.
Mazatlán, Mexico, © 2024 F. Espenak, www.MrEclipse.com*

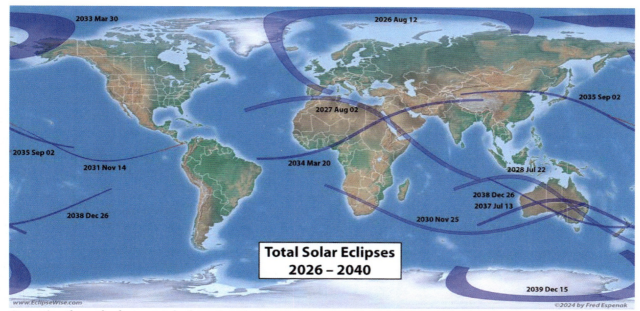

Figure 9 – The path of every total solar eclipse from 2026 through 2040 is plotted on a world map.

10 – Total Solar Eclipses: 2026 – 2040

During the 15-year period 2026 through 2040, Earth will experience 11 total solar eclipses. Examination of the map in Figure 9 shows that only one of them is visible from North America (Alaska on 2033 Mar 30). In contrast, Australia is fortunate to have no less than four total solar eclipses during the same period (2028 July 22, 2030, Nov 25, 2037 July 13, and 2038 Dec 26). Information for each eclipse appears in Table 8.

Table 8 – Data for Total Solar Eclipses

Calendar Date	Time of Greatest Eclipse	Saros Number	Latitude	Longitude	Sun Alt	Path Width km	Duration Totality
2026 Aug 12	17:45:57	126	65°N	25°W	26°	294	02m18s
2027 Aug 02	10:06:41	136	26°N	33°E	82°	258	06m23s
2028 Jul 22	02:55:29	146	16°S	127°E	53°	230	05m10s
2030 Nov 25	06:50:27	133	44°S	71°E	67°	169	03m44s
2031 Nov 14*	21:06:20	143	1°S	138°W	72°	38	01m08s
2033 Mar 30	18:01:25	120	71°N	156°W	11°	781	02m37s
2034 Mar 20	10:17:35	130	16°N	22°E	73°	159	04m09s
2035 Sep 02	01:55:36	145	29°N	158°E	68°	116	02m54s
2037 Jul 13	02:39:25	127	25°S	139°E	43°	201	03m58s
2038 Dec 26	00:58:58	142	40°S	164°E	73°	95	02m18s
2039 Dec 15	16:22:34	152	81°S	173°E	18°	380	01m51s

* The eclipse of 2031 Nov 14 is a Hybrid Eclipse or Annular/Total Eclipse.

Explanation of Column Headings in Table 8

Calendar Date – The UTC date of the eclipse at the instant of Greatest Eclipse
Time of Greatest Eclipse – The UTC Time of Greatest Eclipse
Saros Number – The number of the Saros series that the eclipse belongs to
Latitude – The latitude of the Moon's umbral shadow at instant of Greatest Eclipse
Longitude – The longitude of the Moon's umbral shadow at instant of Greatest Eclipse
Sun Alt – The altitude of the Sun at the instant of Greatest Eclipse
Path Width – The width of the path of totality (km) at the instant of Greatest Eclipse
Duration Totality – The central line duration of totality at the instant of Greatest Eclipse

Total Solar Eclipses from 2026 to 2040

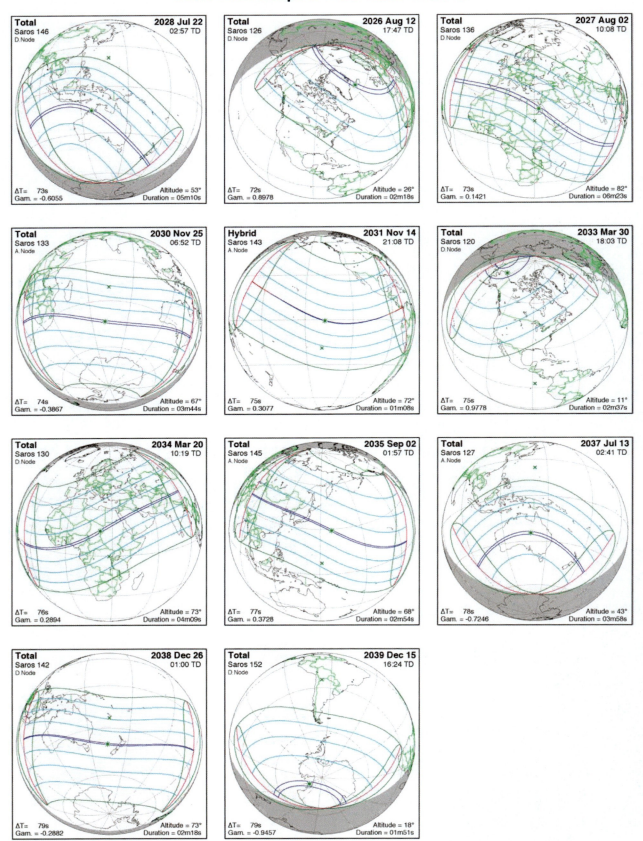

Figure 10 – Total Solar Eclipses from 2026 through 2040.
Courtesy of **Thousand Year Canon of Solar Eclipses 1501 to 2500**, F. Espenak, (2014)

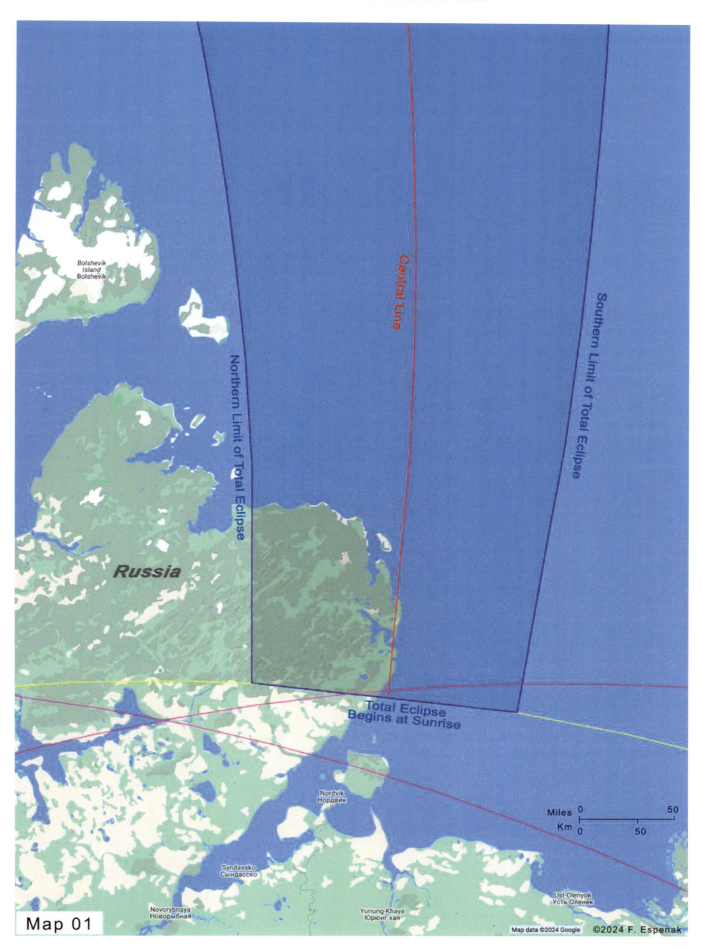

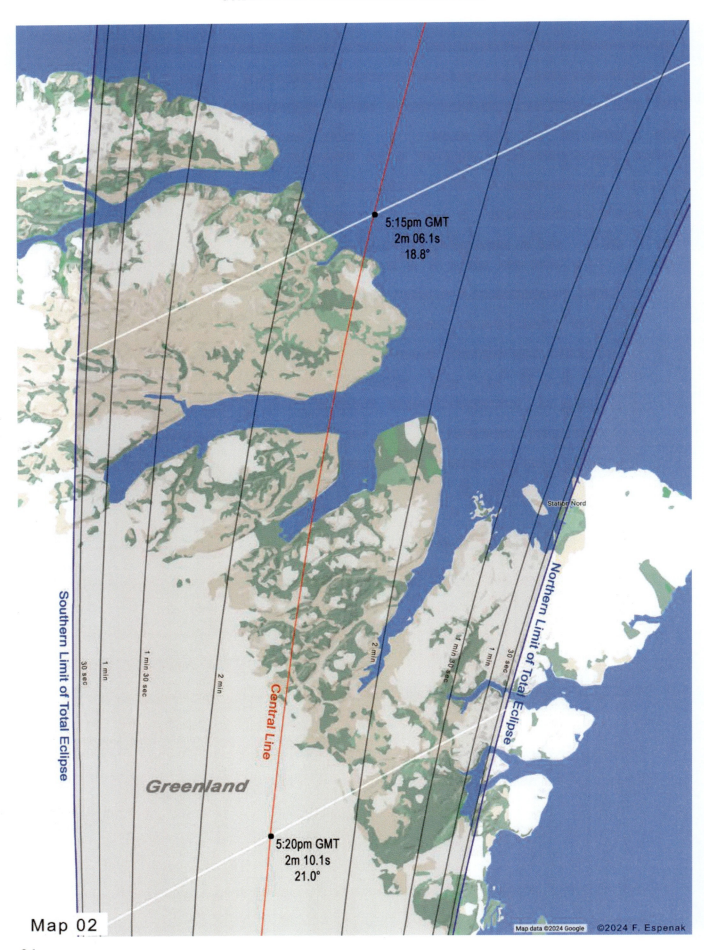

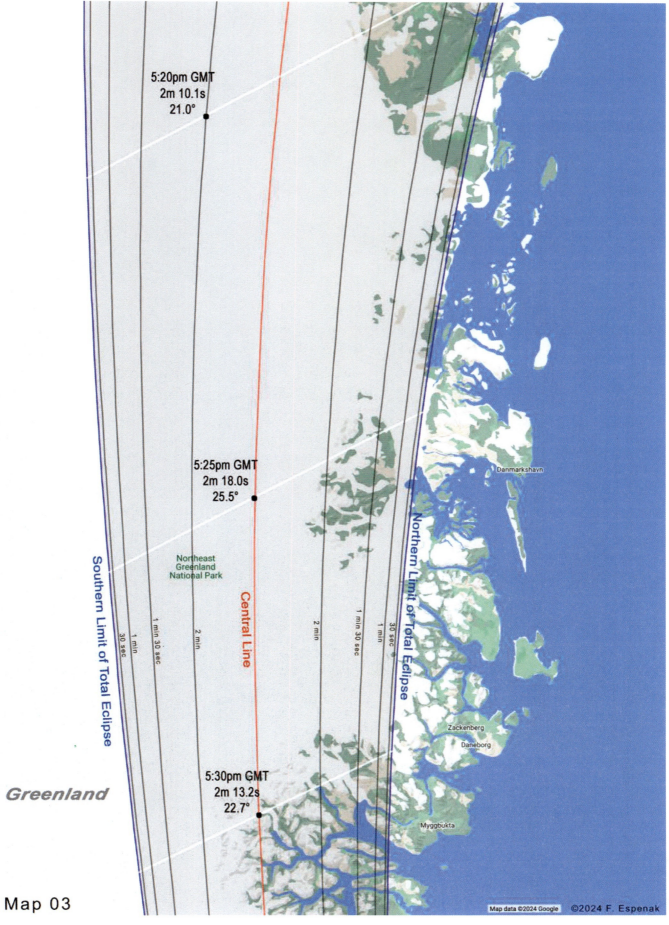

Map 03

Guide for the Total Solar Eclipse of 2026

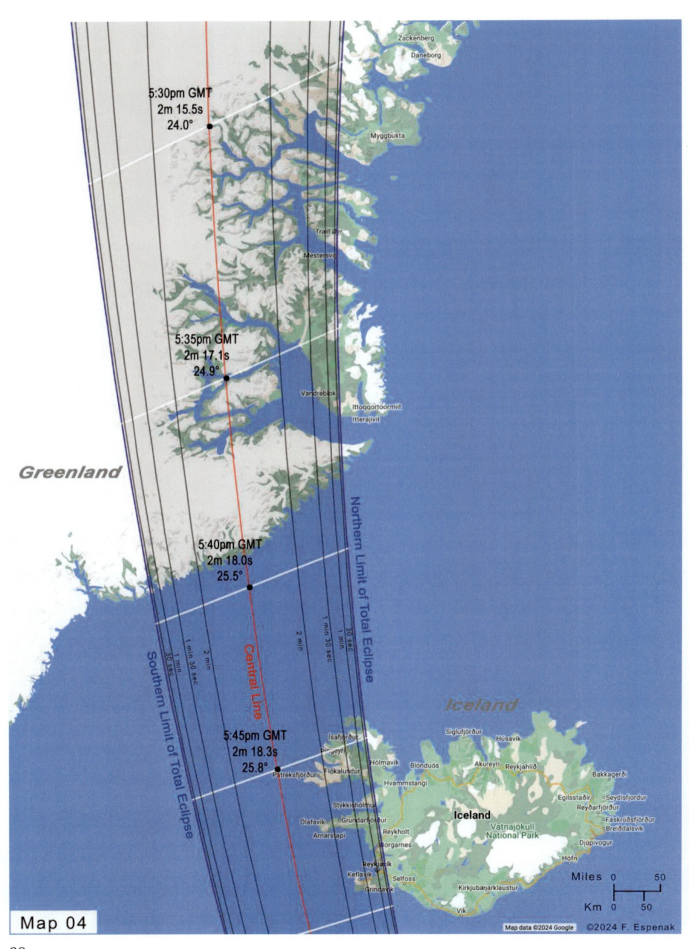

Map 04

Guide for the Total Solar Eclipse of 2026

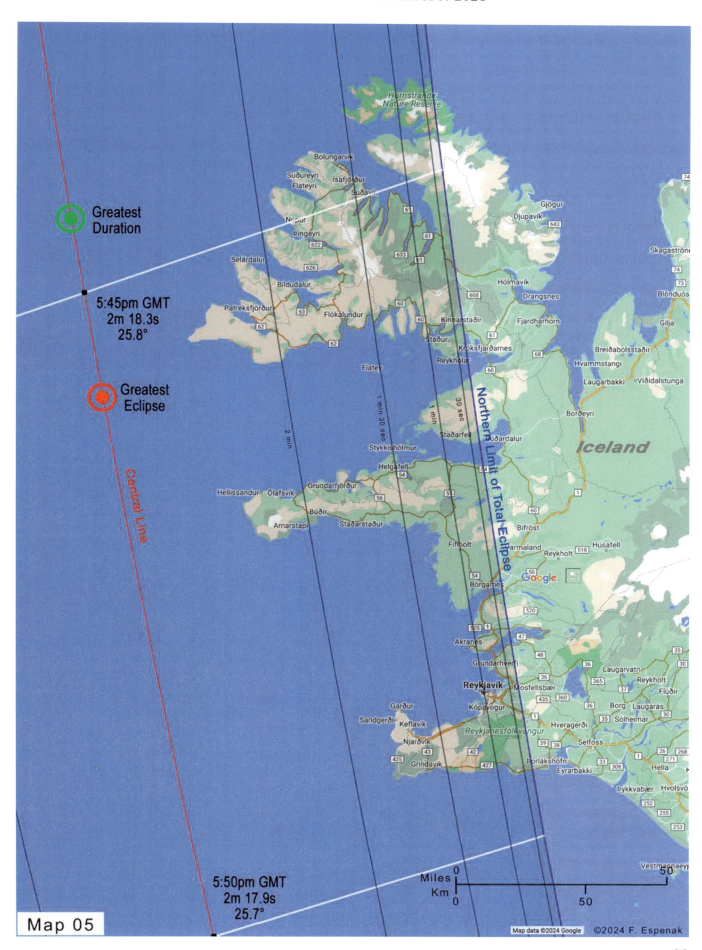

Map 05

Guide for the Total Solar Eclipse of 2026

Map 06

Guide for the Total Solar Eclipse of 2026

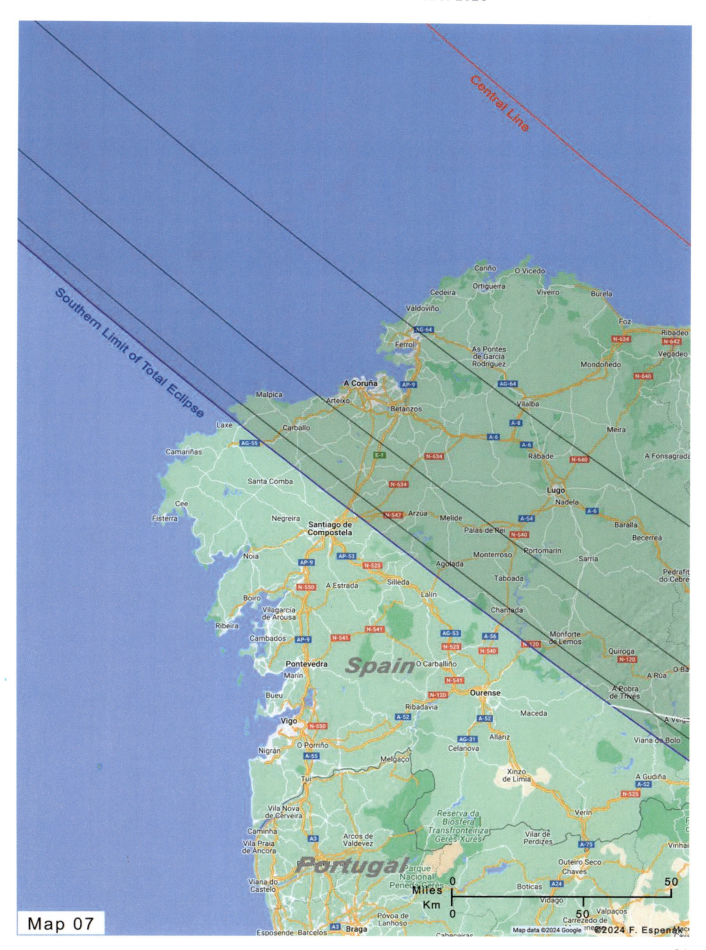

Map 07

Guide for the Total Solar Eclipse of 2026

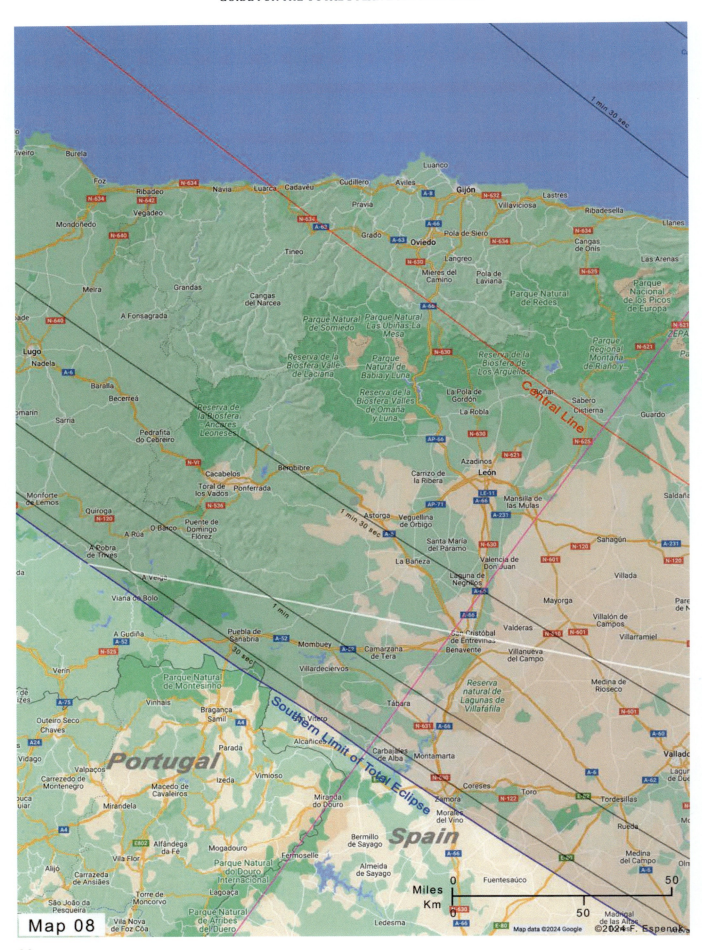

Map 08

Guide for the Total Solar Eclipse of 2026

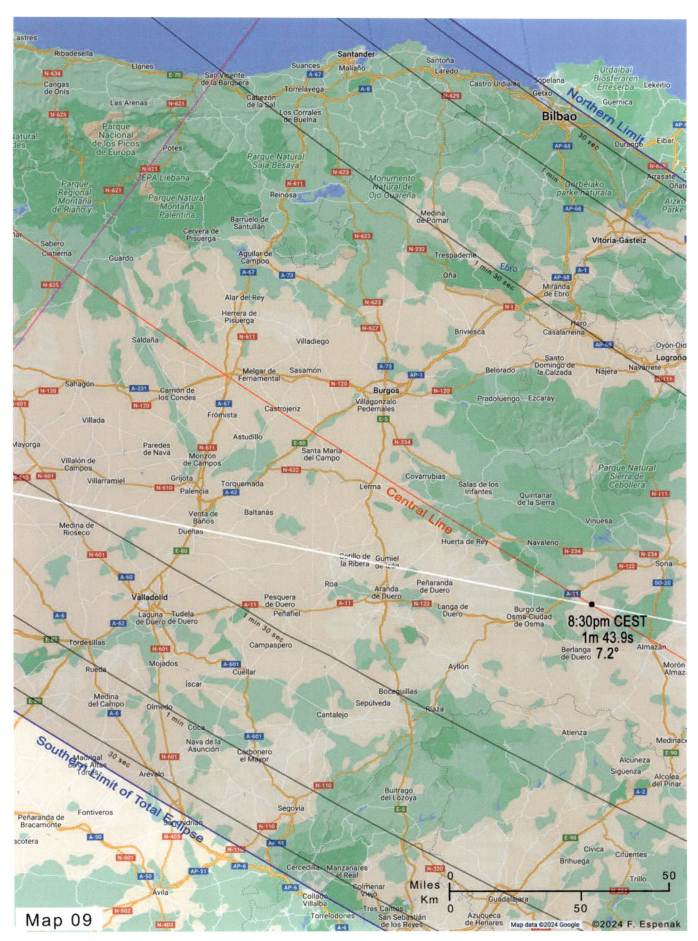

Map 09

Guide for the Total Solar Eclipse of 2026

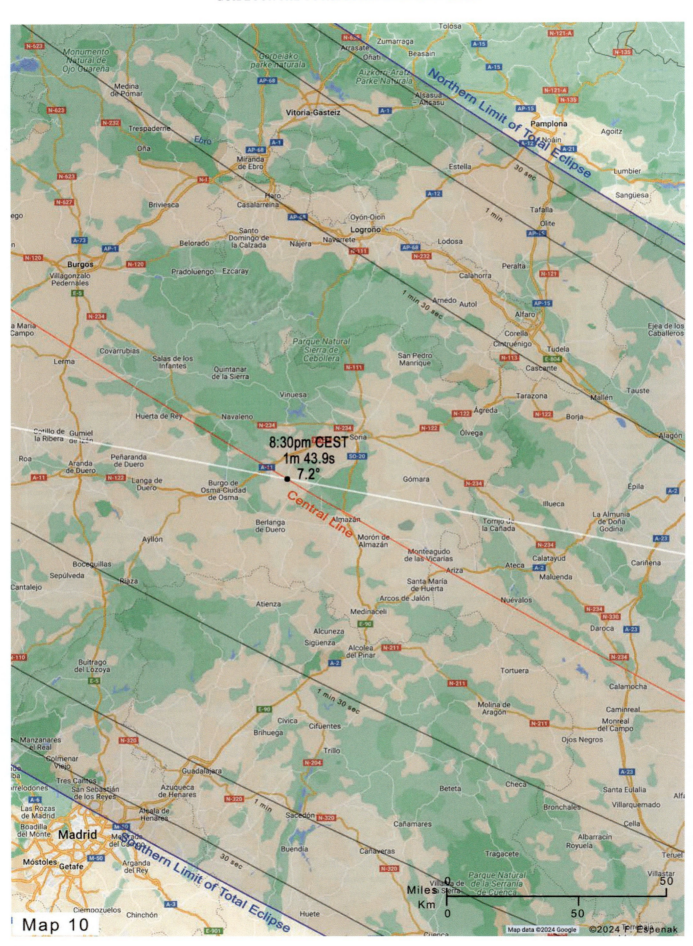

Map 10

Guide for the Total Solar Eclipse of 2026

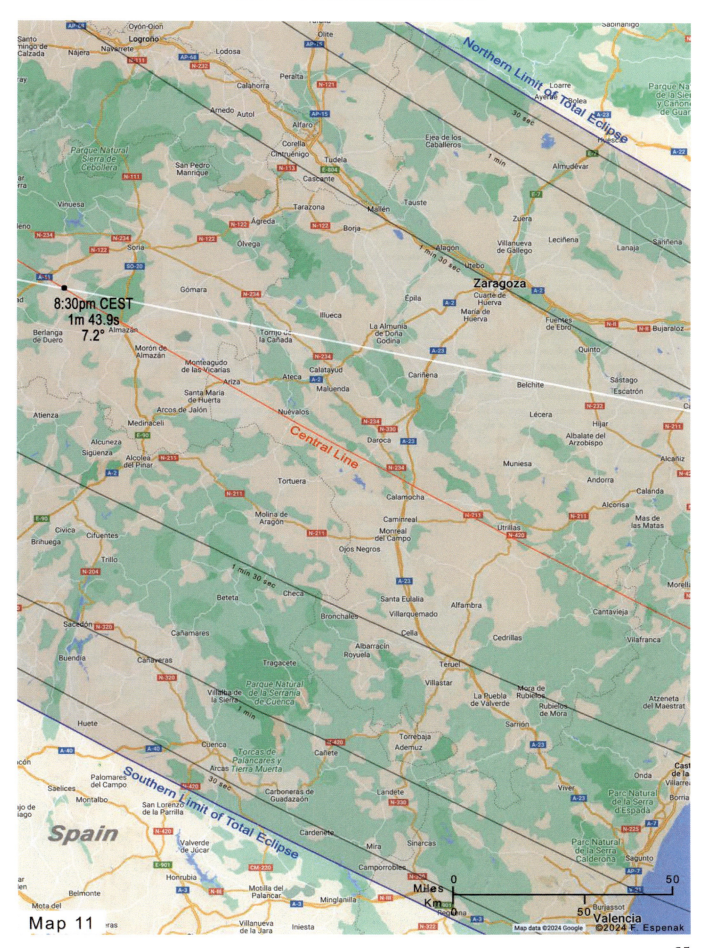

Map 11

Guide for the Total Solar Eclipse of 2026

Map 12

GUIDE FOR THE TOTAL SOLAR ECLIPSE OF 2026

Map 14

AstroPixels Publishing

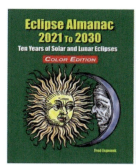 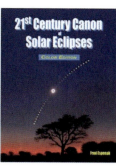 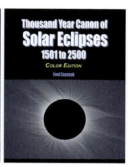 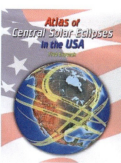 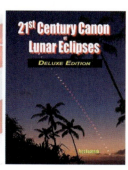

Eclipse Almanac – 2021 to 2030

A concise reference for every eclipse of the Sun and the Moon over a 10-year period. It identifies when and where each of event will be seen with maps for each event. Additional details about each eclipse are included. Other volumes in this series include the decades 2032 to 2040, 2041 to 2050, 2051 to 2060, and 2061 to 2070.

21st Century Canon of Solar Eclipses

The complete guide to every solar eclipse occurring from 2001 tom 2100 (224 eclipses in all). It includes information and maps for all total, annular, hybrid, and partial eclipses. A special world atlas shows detailed full page maps of all central eclipse paths (total, annular and hybrid).

Thousand Year Canon of Solar Eclipses — 1501 to 2500

Contains maps and data for each of the 2,389 solar eclipses occurring over the ten-century period centered on the present era. The maps are arranged twelve to a page at an image scale allowing the determination of eclipse visibility from any location on Earth. A comprehensive catalog lists the essential characteristics of each eclipse.

Atlas of Central Solar Eclipses in the USA

When was the last total eclipse through the USA and when is the next? How often do they happen? What total eclipse tracks passed across the USA during the 17th, 18th, and 19th centuries, etc., and what states did they include? And how often is a total solar eclipse visible from each of the 50 states? The Atlas of Central Solar Eclipses in the USA answers all these questions and more with hundreds of maps and tables.

21st Century Canon of Lunar Eclipses

The complete guide to every lunar eclipse occurring from 2001 tom 2100 (228 eclipses in all). It includes information and maps for all total, partial, and penumbral eclipses. The predictions use a new model for Earth's elliptical shadows.

Thousand Year Canon of Lunar Eclipses — 1501 to 2500

Contains maps and data for each of the 2,424 lunar eclipses occurring over the ten-century period centered on the present era. The maps are arranged twelve to a page at an image scale allowing the determination of eclipse visibility from any location on Earth. A comprehensive catalog lists the essential characteristics of each eclipse.

All books are available in two editions: 1) Black and White, and 2) Color.

For more information including sample pages of each, visit:

astropixels.com/pubs/

Printed in Poland
by Amazon Fulfillment
Poland Sp. z o.o., Wrocław